THE POWER OF IDENTITY IN A PLUGGED-IN WORLD

DON PEARSON

pot-boilers

Copyright © 2015 by Don Pearson

www.don-pearson.com

ISBN 978-0-9863012-0-9

All rights reserved. No part of this book may be reproduced in any form or by any means, electronic or mechanical, including photocopying and recording, or by any information storage and retrieval system, except as may be expressly permitted by the publisher.

All scripture quotations, unless otherwise noted, are taken from THE HOLY BIBLE, NEW INTERNATIONAL VERSION®, NIV® Copyright © 1973, 1978, 1984, 2011 by Biblica, Inc.® Used by permission. All rights reserved worldwide.
All emphasis is the author's.

Scripture quotations marked (NLT) are taken from the Holy Bible, New Living Translation, copyright © 1996, 2004, 2007 by Tyndale House Foundation. Used by permission of Tyndale House Publishers, Inc., Carol Stream, Illinois 60188. All rights reserved.

Branding – Apricot Services

Cover design – Sarah Molegraaf

Cover Image – Shutterstock

Concept Edit – Parents of Adolescents Class

Editorial – Alexis DeWeese

Copy Edit – Julie May

Interior layout – Pete Ford

To Matt

Your technology use magnifies your love for Jesus

CONTENTS

00/

Your life is consumed with these symbols. Hit *play*, trigger the song or video, enjoy. We hit buttons all day and enjoy the results without thinking much about the technology that delivers the stuff. Sometimes in all of this, the buttons get lost. We forgot that we're making choices.

Michael uses the buttons. But he never thinks, for example, about the order. Or about what the buttons might symbolize in his deeper life. Michael's secret looks like this: *play & stop.* He doesn't think about the button and therefore, never reflects on why his addiction is so powerful. He remains in a love-hate cycle with pornography. *Play.* Stop. *Play.* Stop. You know all about choices because you make your own. On your device. But once they get embedded into your life, they kind of run like a perpetual motion machine. Like a battery that never needs charging.

Something is driving Michael's problem. And it has very little to do with sexual urges. That little *pause* button? What if he were to stop for a moment and think about one word. *Why?* "Why am I doing this?"

I could tell that the real issue hadn't come up yet. We were sitting at Starbucks and Sarah had already taken the lid off, snapped it back, taken it off, snapped it back, taken it off, snapped it back.

"What's bothering you lately?" I asked, sipping my mocha – with one less pump of chocolate.

She looked at me and *paused* and stopped fiddling with her cup, "I push guys away."

I waited.

"It's like I try so hard but end up with the exact opposite of what I want."

"What do you want?" I asked.

Her head slowly bounced back and forth, like a ping-pong idea was moving around inside it, in slow motion. Then her eyes locked onto mine, "It'd be nice to be noticed once in a while, right?"

Sarah went on to talk about dumb mistakes she'd made with guys. Texting, *kind of* sexting, trying to arrange being at the same parties.

Sarah and Michael's choices are very different. At least on the surface. But the energy that drives them comes from a similar place. Think back to the *play/stop/pause* buttons. We're so busy managing the *play* & *stop* that we seldom take time to hit *pause*. To think deeply. To connect our surface struggles with the deeper source of what's driving them.

PayPal had an ad slogan a while ago: *Pay at the speed of want*. Your wants appear instantly but are driven by forces buried deep inside. The place where God specializes in hanging out.

Ever wonder why God feels so distant? Why He doesn't seem to be as powerful as your struggles? You're not the first to feel this way:

> *Why, LORD, do you stand far off? Why do you hide yourself in times of trouble?* (Psalm 10:1)

What if you stepped into an adventure and began to use *pause*, to connect the dots of why the struggle is so frustrating? What if God was more interested in the deep parts of your life, the ideas and secret wants that drive your surface issues?

This book is about making that connection. Your technology, your relationships, your identity, your deep desires. You already know what happens to your friends who just use technology randomly, without hitting the pause button to think. Their lives become a toxic mess.

When you begin to connect your technology use with the deeper motivations of what's driving it, your view of God will gradually change from being weak and distant to being intimate, personal and powerful.

I don't make the promise. God does.

01/

The Issue is Never the Issue

You're watching a movie and you know you just missed something important. You saw what happened but you need think-time to put the pieces together. Since you have the controller, you hit *pause*...

Same thing with life.

If you float along only using the *play* and *stop* buttons, you'll have a love-hate relationship with your current struggle. You know the drill. *Play* button, feel guilty, *stop* button, pressure, *play*, guilt, *stop*, pressure...repeat...10x.

The power is in the *pause* button.

When you hit *pause*, you're asking a simple but deep question, "What is really going on?"

Why do I do what I do?

Pause button.

Go ahead. Touch it.

Because the issue is never the issue.

Kellyn jumped in the car and glanced at her phone as she back away from the house. 7:17pm. Ignoring the fact that she'd just made several promises to her mom – promises she wasn't able to keep – she forced herself to drive slowly until she turned the corner.

She needed to get to three places before 8:30pm.

Battery?

28%.

She frantically felt around on the floor until she located the charging cord and corrected her swerving while she plugged the phone in.

"Almost there," she texted Emily, just before calling Stephen.

He didn't answer. He never answered. Resisting the urge to slam the phone against the steering wheel, she texted Suzanne, "Tell Stephen I've got to talk with him."

She looked both ways at the four-way stop before rolling into a left turn.

Homework! She remembered, shaking her head.

Mentally, she went through the schedule again. A quick stop at Applebee's to make sure Emily didn't ruin the whole thing. Then the store, to satisfy one of her mother's requests, and then the big one – the party at Stephen's house.

Kellyn is addicted to being the celebrity. Being *on* all the time. It's important for her to show up everywhere, get the pics and keep all the relationships spinning. Later – in her bedroom – she'll craft it together into a flawless image. The social networking world. Grades. Friends. Sports. Job. Parents. A giant pile of stress from which she has to maintain a perfect portrait. She's fighting the gravitational pull of becoming a non-person, of sinking into the hole of invisible. And she barely got her mother to let her go tonight.

What's your story? Is it similar to Kellyn's? Or more like…

Calvin slung his backpack off and let it thump on the floor. He read the note from his mother. *Snack on the counter. Dad's gone 'til Thursday, I'll be home late. Do your homework first, before video games.*

Calvin crunched the note in one hand, "Like Dad is ever home?" he said aloud, his words bouncing off the kitchen tile. "Pretty much the same thing she said yesterday."

Downstairs, he flipped his laptop open while glancing back at his phone. No texts. He opened his school planner and stared at it without reading. Brutal.

Turning his phone over, he got up, closed the curtains and flipped the lights off. Then he fired up the big screen and controller and put the headset on. Calvin, the gamer.

Back to your story. What *play* buttons do you keep pushing:

- Texting, glued to your phone
- Emotional porn – creating pretend situations in your mind
- Reading books that distort your sense of reality
- Masturbation
- Alcohol
- Porn
- Overuse of food
- Overdosing on social media
- Internet surfing

- Video games
- TV
- Pot
- Same-sex attraction
- Hooking up
- _____

Want some good news? The issues – the play buttons you just identified – are not the main issues. Want an adventure? Hit the *pause* button and start thinking.

Overdosing on Facebook – or any other temptation – is like an iceberg where the only part we see is the exciting small tip of ice above the waterline. When we're in the grip of temptation it feels almost impossible to hit the *stop* button, right? But what if you hit the *pause* button instead? For just a moment? What you learn could change your life.

Check out this diagram...

Iceberg

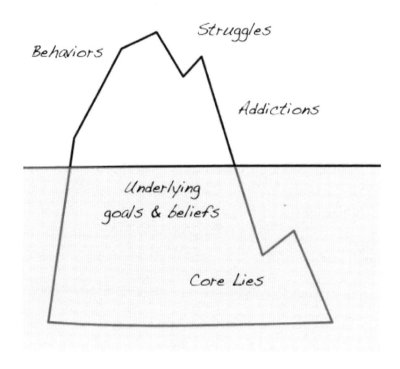

The biggest part of the iceberg is underwater, totally invisible. And until you look at that part of your life, you'll never crack the code of what's going on topside.

Let's hit the *pause* button in Calvin's life and look at him through the image of the iceberg.

The stuff above the waterline is easy. Calvin's in a video game rush. He never thinks about why he's doing it and that's the reason he's addicted. Right? He's stressed out at times and he just needs some down time. Video games provide it for him. End of story. Except for one little problem.

It doesn't work.

It doesn't satisfy. If it did, he wouldn't need more, more, more.

It's only a symptom, not the real issue. But it's consuming large parts of Calvin's life.

So back to the *pause* button.

What's really going on? What's the huge stuff, below the water line, that is pushing upward and providing the fuel for his gaming? Calvin doesn't know it yet, but he's on the edge of a fantastic adventure – an inward journey somewhat like a video game with enemies, ammunition, fear, allies, surprise, shock and awe.

When Calvin hits the *pause* button and answers a few basic questions, the under-the-water part of the iceberg begins to come into focus:

1. What is going on in his life?
   - The pressure of homework and grades never stop
   - The house is usually empty
   - His mom is always telling him what to do
   - His dad basically ignores him
   - He needs to keep *working* his friendships or they disappear

2. What is his view of himself based on all that stuff?

- He's not interesting enough to get his dad's attention
- Although he has a group of 'friends,' he doesn't get many invites
- He can never please his mom
- His homework is never done

3. What does gaming promise him?

- He is powerful
- He has what it takes
- People notice and respect him

Notice the third section – the deepest part of Calvin has *nothing to do with video games*. Instead, the magnetic pull of spending hours and hours in the dark is supercharged because of the deep stuff – the issues that are under the water.

Your life is like this. Although the surface issues in Kellyn's or Calvin's lives may be different than yours, the core problem is strangely similar because we are all human. Check out this verse,

No temptation has overtaken you except what is common to mankind. (I Corinthians 10:13)

And since this is deeply private and personal, it's only going to involve two people. You and Jesus.

Yes, I know. That may seem weird and impossible. Where do you find Jesus anyway? In a freaky church building, hanging on a wall with a diaper on while blood drips down from His head?

Jesus put you together. And He's the only one who can be your companion on the inward adventure. At times He walks beside you and points to a few things you weren't aware of. At other times, He's a step ahead of you and turns around to make sure you're still coming. On one occasion, He'll push. On another, He'll pull.

DRILLING TO THE CORE

A four-step approach is often helpful in linking surface struggles with underlying, core issues. Warning! This isn't just a five-minute exercise – it's a whole new way of thinking about life.

– *Step One* – Choose a struggle in your life and begin to ask God for insight. How is it linked to a deeper issue?

– *Step Two* – Be honest about your pain. What are the big events in your past that hurt you? How did these events help shape your identity? What lies did you start telling yourself? If you could narrow the lie down to a negative word to describe your self-image, what would it be? (for example: unwanted, invisible, fat, loser, second string).

– *Step Three* – What "secret promises" (often subconscious) did you make following these hurtful events? We make the biggest decisions of our lives following painful experiences. Calvin's hurtful relationship was with his father. So he told himself, "I will never let someone get close to me again." Kellyn's secret promise was "I will always surround myself with plenty of activity and noise." Which of the following options best describes you?

- **❿** I'm trying to prevent something from happening (Calvin)
- **❿** I'm trying to make something happen (Kellyn)

– *Step Four* – Tell God you're sorry. And as you do this, realize that He is more heartbroken about the lies and secret promises you've told yourself (in steps two and three) than He ever was about your surface struggle (step one). The deep, dark rebellion at the core of every human sin is the decision to handle life our own way.

Sometimes it helps to see it in someone else's life first. So let's check out two big events in King David's life. Yes, that David. One of the greatest guys in the Bible:

> *In the spring, at the time when kings go off to war, David sent Joab out with the king's men and the whole Israelite army. They destroyed the Ammonites and besieged Rabbah. But David remained in Jerusalem.*
>
> *One evening David got up from his bed and walked around on the roof of the palace. From the roof he saw a woman bathing. The woman was very beautiful, and David sent someone to find out about her. The man said, "She is Bathsheba, the daughter of Eliam and the wife of Uriah the Hittite." Then David sent messengers to get her. She came to him, and he slept with her. (Now she was purifying herself from her monthly uncleanness.) Then she went back home. The woman conceived and sent word to David, saying, "I am pregnant."* (II Samuel 11:1-5)

If you speed-read this story, you probably think it's about sex. But if you use the *pause* button, you'll realize how deep the human heart really is.

Just like Kellyn and Calvin, King David blew through several stop signs without even considering the *pause* button. Check out the warning signs that he missed:

- ❶ Kings go to war in the spring.
- ❶ This spring, David didn't go, but remained in Jerusalem.

- David's identity was closely linked to his military power.
- He couldn't sleep, got up from his bed, and paced the roof of the palace.
- The palace was full of wives and concubines – lots of legitimate sexual options for a king at that time in history.
- But instead, he gazed beyond the palace and saw a beautiful woman bathing.
- He inquired about her and found out that she was married and that she was somebody's daughter.
- He ignored those two inconvenient facts.
- He sent for her.
- He had sex with her.

Don't miss the tension in the story. Real men fight. Wimps stay home. David's male identity was under fire and the surface issue became a sinful attempt to touch a core issue in his life. In other words, he tried connecting the deep problem of feeling insecure (I should be fighting in the war, maybe I don't have the ability anymore) with a surface solution (I still have power).

God was there for him. God was available. God was willing to discuss the deep parts of David's insecurity and remind him of the truth. How did David miss the connection? Why do we?

Here's another story in David's life. But this time he *paused*:

After Saul returned from pursuing the Philistines, he was told, "David is in the Desert of En Gedi." So Saul took three thousand able young men from all Israel and set out to look for David and his men near the Crags of the Wild Goats.

He came to the sheep pens along the way; a cave was there, and Saul went in to relieve himself. David and his men were far back in the cave. The men said, "This is the day the Lord spoke of when he said to you, 'I will give your enemy into your hands for you to deal with as you wish.'" Then David crept up unnoticed and cut off a corner of Saul's robe.

Afterward, David was conscience-stricken for having cut off a corner of his robe. He said to his men, "The Lord forbid that I should do such a thing to my master, the Lord's anointed, or lay my hand on him; for he is the anointed of the Lord." With these words David sharply rebuked his men and did not allow them to attack Saul. And Saul left the cave and went his way.

Then David went out of the cave and called out to Saul, "My lord the king!" When Saul looked behind him, David bowed down and prostrated himself with his face to the ground. He said to Saul, "Why do you listen when men say, 'David is bent on harming you'? This day you have seen with your own eyes how the Lord delivered you into my hands in the cave. Some urged me to kill you, but I spared you; I said, 'I will not lay my hand on my lord, because he is the Lord's anointed.' See, my father, look at this piece of your robe in my hand! I cut off the corner of your robe but did not kill you. See that there is nothing in my hand to indicate that I am guilty of wrongdoing or rebellion. I have not wronged you, but you are hunting me down to take my life. May the Lord judge between you and me. And may the Lord avenge the wrongs you have done to me, but my hand will not touch you." (I Samuel 24:1-12)

Pause button. When we look at the real issues in David's life, it's fantastic that he was able to handle this event correctly:

- As a young man, he killed Goliath. But instead of being honored by King Saul, the king became jealous.
- This jealousy caused King Saul to hunt David for 14 years. He chased David and kept him on the run for what must have seemed like 'forever.' To kill him.
- And then the event happens. Saul enters the very cave where David and his warriors are hiding out. Saul takes off his robe, puts it on the ground and steps away to pee.
- David's men whisper, urging him to kill Saul. They even quote something that God had told David. They say, "Here he is. God provided this chance."

- ❶ David *pauses* and even feels guilty about cutting a piece of the king's robe.
- ❶ David *pauses* and remembers. "God is God and I'm not. It's His job to take the life of an anointed king, not mine."

Pause button! David used it and it became one of the defining moments of his life, where trusting God and trusting God's timing ultimately propelled him to the throne as king. Why? Because the battle is always in the thought world. Imagine if David had been able to *pause* and think when he was restless that one night, when he took someone else's wife.

Notice how small God is in your life when it comes to surface-level struggles. It seems like you pray and pray but God doesn't fix the issue. God is small. He's distant and doesn't seem to be a big part of your life. The reason he seems small is because the issue isn't the issue. It's like begging God to fix the bumper when the car stopped running.

Hit *pause*. Look underneath the issue. Avoid the boring life and choose an adventure with God instead. Like David, your real battle will always live in the world of thoughts. That's when God becomes big. That's how He comes near.

It is God's will...that each of you should learn to control your own body in a way that is holy and honorable, not in passionate lust like the pagans, who do not know God; (I Thessalonians 4:3-5)

God understands that it's a process. No magic. Doesn't happen overnight. Not a fast-food drive-through option. It's about learning to control your own body.

Kellyn is flying through life and ignoring the process.

Calvin is flying through battle scenes and ignoring the process.

Both are only using *play*.

Not you.

You can enter the portal by hitting *pause* and reflecting on why you do what you do.

ⵙⵙⵙⵙⵙⵙⵙⵙⵙⵙⵙⵙⵙⵙⵙⵙⵙⵙⵙⵙⵙⵙⵙⵙⵙⵙⵙⵙⵙ

POSSIBILITIES

– What is a recurring struggle in your life where you'd like to see God's power take over? (use a secret code word if necessary).

– Name one or two big events in your past that really hurt you.

– What lies did you start telling yourself as a result?

– Pick a negative word to describe your lie.

– When it comes to your struggle, are you:

  ❶ Trying to prevent something from happening?
  ❶ Trying to make something happen?

– Tell God you're sorry as you focus on the deeper part, the part that drives your struggle.

02/

Technology & Identity

Kaitlyn's first clue that Grandpa was coming to dinner occurred at 3:23pm. She walked into the kitchen, saw the chair in the middle of the room, saw the sticky note from Mom.

"Seriously?" she yelled to an empty house. "Just what I need!"

Grandpa was awesome but he always asked. Always. He just couldn't leave it alone. It was like he took it personally. An affront to how he viewed things.

"You dating anyone yet Kaitlyn?" That's how it would come out tonight. At just the right time, between bites, when there was a pause in conversation. Mom's eyes would dart to Dad. Dad would look at his plate before saying something stupid like, "I told her, she's not dating until she's 21."

Lame.

"I just can't understand it." Grandpa would continue, "Someone as smart and beautiful as you are? What's wrong with boys these days? Heck, in my time, you would have had to fight them off with a stick."

Ultimately, she'd have to respond to the awkward topic.

You dating anyone yet Kaitlyn?

No Grandpa, I'm not dating because a technology wave of unprecedented speed and power has overwhelmed every guy I know. It's plunged them into the dark sector of virtual girls. Pretend girls penetrate male bedrooms, classrooms and locker rooms with invasive pics and manipulative ploys for attention leaving guys further rocked. Landing on the age-old problem of male passivity and female controllingness, technology has simply magnified the issue and imprisoned me in the insatiable land of social networking. It's where I fabricate my own version of pretend and measure my own power to get noticed in a formula that promises way more than it delivers.

This response seemed a bit heavy to Kaitlyn, so she decided to answer Grandpa's question with short and simple. "Nope. Not dating. New topic please."

Across town Drew was having problems of his own. He'd just learned that Brad had come out. On Twitter. Drew had an account but rarely used it. So he'd gotten the news later than most of his friends. But he was closer to Brad than those who'd just told him. At least it used to be that way. What would happen now? In two hours they'd be together at rehearsal and he needed to say something.

Drew stared out the window at bare trees, where a few stubborn leaves still clung to branches.

Back to Kaitlyn, Grandpa and Dad. Grandpa is disconnected and doesn't understand how teenagers do relationships. Dad doesn't get it either and is still trying to protect her from getting too close too soon with a guy. He thinks that the only way to get in trouble is through a dating relationship.

Both men are expressing concerns from their own generation but they're missing the real Kaitlyn. Her problems are different, something that neither of them fully understand and technology has a lot to do with the disconnect.

What about you? Technology comes to you and says, "I want to be your manager. I'll arrange all your relationships, your fantasies and your intimacies. I promise to satisfy your deepest longings."

How are you going to connect to it and through it? Will you use it or will it end up using you?

Some date, most don't.

In her private life Kaitlyn views porn, wondering what it's like to be desirable to guys. She's secretly told a friend about it and both girls admitted to experimenting with masturbation from time to time. Like most kids her age, Kaitlyn lives in a world where sexual images, sexual language and sexual posturing are coming at her like a machine gun. When she finally escapes the lunchroom conversation and the issues in the hallway, she has to listen to the teacher bring up the same-sex-attraction issue. Just like he did yesterday.

Notice how the Bible cuts through our current trends and struggles:

*Be very careful, then, how you live – not as unwise
but as wise, making the most of every opportunity,
because the days are evil.* (Ephesians 5:15-16)

iConnect is about harnessing the expansive nature of technology because unless you do, it'll leave a toxic mess in your life. Most of the hot-button issues, most of the challenges in your life, most of what you care about are somehow related to two things: 1) your masculine or feminine identity and 2) how technology wants to be your manager.

As you connect the dots, you'll discover that you're living in the most explosive time in human history. And it all revolves around how information is stored and distributed through technology. How communication is carried. But if your life is only about gathering and rearranging information – through your device – you'll miss out on the biggest piece. Relating. Real relating.

When you were little, you held the pencil awkwardly in your chubby little hand and worked out addition problems. 2 + 2 = ___. When you got older, you entered the realm of multiplication. Numbers got bigger in a hurry. And then you heard about exponents. Serious power. Serious numbers.

That device in your hand? It's power and it's exponential.

Such power can be harnessed to grow your relationship with God and partner with Him to change the world. It can also be used to deeply relate with friends. Technology will be inseparable from your career path and it'll likely have a huge impact on how you journey towards marriage. That's the nature of it. It's expansive. But how you use it will make all the difference in the world. And you can't figure out how to use it without understanding the deeper parts of what it means to be a man or woman. Identity issues have a lot to do with how you use your technology. They're connected.

ⅢⅢⅢⅢⅢⅢⅢⅢⅢⅢⅢⅢⅢⅢⅢⅢⅢⅢⅢⅢⅢⅢⅢ

POSSIBILITIES

- What are some areas of weakness in your life that technology magnifies?

1.

2.

- How are you using technology in a positive way?

1.

2.

03/

Emotional and Visual Pornography

You are going to have a relationship with pornography the rest of your life. It might be something you struggle with, something a friend needs help with or something you despise, but it's now embedded in the fabric of our culture and we can't escape it. It's invasive.

We watch YouTube for the latest music videos and to figure out how to fix stuff. Google and Wiki are close friends when doing homework. You get your hair cut in a place where pictures line the walls. You walk through malls, stand in line in the grocery store, watch movies and scan Facebook. It's invasive.

Here are the two ways it comes at us:

Emotional Pornography – This gets developed in your heart and your mind. It might come through reading or you might just create it while sitting in a boring class. You build out a pretend situation and go looking for a love that does not exist. It's a feeling thing like a dream or scene that keeps repeating in your mind because it feels good. You imagine a cast of characters, usually you and someone else, and you tell that person what to do and they do it.

Visual Pornography – This starts by looking at images, and then gets further developed in your mind. It involves the pretend world and is also built around a love that doesn't exist. It can be as simple as something on YouTube or an image that is more obvious. It's similar to emotional pornography but your imagination is further enticed by images and you're the director of the movie. You give assignments and the images do what you tell them to do.

Hit the *pause* button a minute. Why did *Twilight*, *Hunger Games*, *Divergent* and *Fifty Shades of Grey* become runaway bestsellers? Why is visual pornography a $100 billion dollar industry?

The answers live within every human being. Technology simply links the two together – our human condition and a shortcut to getting what we think we need. Technology acts as our agent and manager, offering us options for our deep loneliness. And because of its power, technology magnifies the issues.

Brianna thinks about Curtis several times a day, imagining upcoming events and how they might play out. It's her favorite way to fall asleep at night, where some unexpected – and if she's honest, unrealistic – situation occurs so that they find themselves alone. In real life, Curtis is polite and texts her back. Sometimes. But to pretend that anything exists between them is imagining a love that doesn't exist.

We'll call this emotional pornography or pE for short.

Curtis is grabbed by the other kind of porn. He doesn't often think of people he knows, but scans lots and lots of images instead. Might be YouTube surfing, might be Facebook stalking, might be the darker stuff. He's in the hunt and to be honest, he's blending pV (visual porn) with pE (emotional porn). Like Brianna, he's searching for a type of love that doesn't exist in reality.

We already know that something deeper fuels our behavior, right? Brianna doesn't know it but she's terrified of going unnoticed, of being invisible. So she plays pretend games in her mind. Curtis is also blind to the real force behind his struggle of sexual temptation. In most of his relationships he feels unwanted, and porn offers a substitute lie at the core level of life.

One of my favorite C.S. Lewis stories is about the kid playing in the mud puddle. Someone tries to pick him up, telling him that they're going to take him to the ocean so he can play in the sand and in the waves, but he pitches a royal fit. Starts screaming because he has to leave his mud puddle.

Here's the point.

The kid has never seen the ocean. So the word doesn't mean anything to him and he goes on thinking about the mud. His desires are too small.

Pornography shows us that we don't have a very good vision of God. We prefer the mud puddle because we've never seen dolphins, sandcastles, waves, seashells, boardwalks, great ships, sea turtles, sharks, crabs and a million other things. Mud is what we want.

Enter technology.

When technology meets up with our areas of vulnerability (deep issues), it's like putting something under the microscope. It gets magnified big-time. That's why you have to link: 1) your feminine or masculine tendencies with 2) how you use your technology.

While I'm writing this book, the world is obsessing over Ebola – the insatiable killing machine that can liquefy a body, making it vomit blood and brain sludge. Like Ebola, pornography is a hot virus, but of infinitely more sinister intent. It searches for a host. Both pE and pV seek to bleed out the soul of every host they enter and it's interesting that they don't have any power until they infiltrate.

You're the host.

As the virus begins to replicate itself in the host, the host turns into...it.

The hot agent virus attacks the deepest part of our being, bleeding out real love and all that it means to be human, and replacing it with pretend. With empty. With fantasy. It seeks to replicate itself. A love that does not exist. To look at a nude picture and begin to separate the person from the image – telling it what to do, how to behave – is to enter a pretend world where love does not exist. In similar fashion, attaching my emotions and romantic notions to a make-believe story is nearly identical. It doesn't exist.

When dealing with infectious disease, doctors often encounter superstition in remote villages. At the onset of a killer-virus like Ebola, fear causes many villagers to cut trees across the two-tracks and block the roads that could let help through for their community. For some reason, fear of the outside world causes them to shut down and isolate. Like a reverse quarantine. Like secrecy.

pV and pE also grow through the secret life. Cut off from outside help, the hot agent flourishes in the task of replicating itself. Until the host self-destructs and every last ounce of humanity bleeds out.

When a new outbreak occurs, what do the infectious disease experts do? Where do they go? They rush off to find 'patient zero,' right? Where the hot agent originated. Because if they can locate patient zero, they may be able to find the source.

The source of pornography is usually a person's core lie. Once the hot agent attaches itself to the lie, you begin to see it replicating itself like a level 4 virus. It feeds off the lie, multiplying its hold and power on the individual until it reduces the human host to a 'love that doesn't exist.'

Above all else, guard your heart, for everything you do flows from it. (Proverbs 4:23)

Ebola and other level 4 viruses require serious measures. All-out war. Shouldn't pV and its cousin pE require the same intensity?

❶❶❶❶❶❶❶❶❶❶❶❶❶❶❶❶❶❶❶❶❶❶❶❶❶❶❶❶

POSSIBILITIES:

- Which hot agent are you most susceptible to? pE or pV?

- In what way does technology magnify the problem?

- Where are you most in danger of technology's reach into your life?

- Think of someone who is ahead of you in working through the struggle.

- How big of a risk would it be to talk with them, break the secrecy and ask for help?

04/

Blow It Up

Logan was 13 years old when he made the phone call that would change his life forever. Like you, he was simply trying to figure life out in the middle of a very hard day. He lived on a ranch in Nebraska and part of his responsibilities included taking care of animals.

One particular day he lost the battle of trying to nurse a calf back to health. The small cow had been behind schedule since birth and Logan had grown very attached, thinking he could save it. But it wasn't to be. On that day, Logan would take a gun and put the calf out of its misery.

What happened next is an example of how technology expands everything and how you can begin to think differently, how you can start changing the world.

Logan had been in the habit of calling a local radio station and sharing his thoughts about Jesus and his relationship with Him. On this day, Logan made another call. Through choked-up emotions, he related the story about the calf and his broken heart. He drew out some similarities about what it must have felt like for God to lose His son Jesus, for the sin of the world. He cried.

The radio station was so impacted by his story that they decided to air the call. A church heard about it and put it in video form. People started emailing each other. And within 3 weeks, the station's website crashed because it couldn't handle the request from six million people to hear the recording.

Six million.

A home-school boy, an isolated ranch, a simple calling to take care of animals. The story of God's love went viral. Logan blew it up.

Everyone has an opinion about technology. Some say it's evil, as if it has a life of its own. Others comment on it being neutral. Like anything, it can be used for good or for bad. Some of you live in a strict home while some of you can do anything you want. The same is true for your friends. Some of their homes have more rules than yours, and some have no limitations. Families are all so different when it comes to technology.

The mechanic slammed the hood of the truck, "Better tow it to the dealer, no idea what's wrong..."

The kid was fiddling with his phone and spreading his fingers to enlarge the words. "This make sense to you?" he asked, handing the phone to the mechanic.

Greasy fingers cradled the device and a wrinkled expression hit the man's face before he returned the phone and crawled under the driver's door. The sheet metal trapped a muffled expression before the kid heard, "Try it now."

The key turned, the ignition fired, the exhaust pipe hummed its mellow tune.

The mechanic scooted out and stood and knocked the gravel off his pants, then gave the kid a confused stare.

TECHNOLOGY – THE SPEED OF CHANGE

When your parents were your age, lots of teenagers dated. And when they dated, kids got too close, too soon. They touched each other too much. By the time these kids became parents and had teenagers of their own, they began to parent from their own history. After all, that's how people raise the next generation. They learn from their mistakes and try very hard to ensure that the next generation doesn't repeat them. They say things like, "You're not dating until you're 16," or "You should buy a purity ring," or, "Make sure you know where your physical boundaries are."

In the old days, these methods were often helpful. Time moved slow enough to allow parents to learn from mistakes. A generation learned and then parented from that knowledge.

Doesn't work that way now.

Technology is changing life way too fast.

Most sexual mistakes made by students today happen outside a dating relationship, not within one. Technology has a drop-down menu called "stupid sex" and it's longer than your screen.

But there are technology wins. Great wins. Like the story of Logan. Your opportunities to impact someone's life for Jesus are increasing every few months because of the speed of technology change.

Have you ever been around someone who has a case of vertigo? They're dizzy. They have their eyes closed and their hands out. They get on the ground because their whole world is spinning. Technology has created a type of vertigo. Change. Spinning. Dizzy.

Pretend for a moment that one of your parents offers to buy you the device of your choice. What would it be? Where does your mind go?

Hit *pause*.

Allow yourself to ask a different set of questions. How could this device be used to expand how God is meeting me in my struggle? Logan cried. You cry – at least inwardly. What do you cry about and how does Jesus want to invade that spot? The good thing about technology and the not-so-good thing is that it expands. It magnifies. It blows things up. By hitting the *pause* button and thinking first, you'll be able to use your technology to impact the world rather than letting it use you.

When Alexander Graham Bell invented the telephone he said,

I don't think I'm exaggerating the possibilities of this invention when I tell you that it is my firm belief...that one day, there will be a telephone in every major town in America.

Seriously? A telephone in every major town in America? And then the guys landed on the moon, right? And 'telephoned' home, "That's one small step for man; one giant leap for mankind."

So far so good. Progress through advancing technology.

But then we got freaked out.

The guy standing on top of the African hill was dressed like a Zulu Warrior, in full costume, except that he was on the front cover of magazines all across the world. The only thing out of place was the iPhone in his left hand. The caption underneath the picture said,

This man holds more information than did the last president of the United States at the height of his presidency.

We now give those same devices to 12-year-olds. Or 10-year-olds.

Freaky weird.

Freaky fast.

What it boils down to is that the people just ahead of you can no longer advise you from their own experience. If you ask, "Hey mom, how many texts did you average per month when you were 16?" you'll get the look, right? Well, try to imagine your future daughter asking, "Hey Mom, how many streaming watts did you render, how much sensory input did you allow your avatar when virtually hammocking in the Bahamas?"

By the time an experience turns into wisdom, it's no longer helpful because it's outdated. The speed of technology has changed everything.

TECHNOLOGY IS NOT NEUTRAL

We can use technology in both positive and destructive ways. No one argues that. But that doesn't mean it's neutral. Let's pretend that my job is to create new video games. So I put you in a chair, hook up wires to your head and map the blood flow in your brain. I test different things out on you to see what works. What part of my trial video game fires the most pleasure in your brain?

Once I know the answer, I have you.

I offer you a free trial. "Here you go, play this game all you want." But in the game, I deliver you to the edge of mind-blowing pleasure and stop. I build a firewall into the game. In order to continue to the next level, you need to pay me $5.

I'm rich.

Why?

Because technology isn't neutral. It triggers a dopamine release. Dopamine is a chemical that's released in your brain and it's related to behavior and rewards. We like the stuff because it magnifies and enhances certain things. It feels awesome.

You, your friends, your parents, your teachers. They all have a preferred dopamine drip. Maybe it's video games. Maybe Twitter. The business person feels good once they kill off all the emails. The sports fan watches a football game on a giant screen with two other devices feeding him relevant stats. Dopamine. We all like it. Technology is not neutral.

TECHNOLOGY MAGNIFIES MALE & FEMALE TENDENCIES

The word 'sex' got taken hostage by physical and romantic notions in the 1900s. Before then, it meant male or female. A newborn baby prompted the question, "What sex is it?" We now use the word gender, which seems to be a strange word. All it means is that being a guy or a girl is a heck of a lot deeper than your body parts. It means that your soul is fashioned in either a female or male mold. It's your sexual identity. Hold that thought and we'll discuss it more in separate chapters on guys and girls.

When you *pause*, you will be amazed at how technology magnifies our deep parts. You can learn a ton about your friends and their core identity by watching how they use their technology.

TECHNOLOGY CAN MAGNIFY THE DISCONNECT

When Gabby discovered that she'd been excluded from the latest round of texting and wasn't invited to the beach, she took it to a new level. She developed a better option, texted all her friends *except* the girl who'd excluded her. New plan. Better plan.

Or think about a meal with five people at it. If one of the members is constantly on their device, how does that affect the discussion? Who's in and who's out? The disconnect is obvious.

DIGITAL TECHNOLOGY AS THE NEW METHOD OF

COMMUNICATION

Get this. Radical shifts occur whenever a new means of storing and distributing information occurs. The weird thing is that there have only been four major methods in human history. Like being strapped to a rocket, you're at the front end of the fourth. Here they are:

- *Oral* (Creation-1500 A.D.) - People stored and distributed information by talking face to face. It was the only option.

– *Print* (1500–1950) – When the printing press got invented (technology), we could communicate with someone we'd never seen, with people we didn't know. You and I are communicating right now because you're reading my book.

– *TV* (1950–2010) – This really messed with people. It could broadcast random images and cut through the nice, neat stories that were printed.

– *Digital Interactive* (2010–?) – Words, sounds, pics and data are all interactive and group-powered, without boundaries and all shaken together. Powerful.

Want to think about something strange? Thomas has a grandmother who gets most of her information by reading. His father, on the other hand, is a TV guy and is glued to Fox News. Thomas does neither. He's on his device.

Three generations in one house. Three different ways to trade information.

But you, Daniel, keep this prophecy a secret; seal up the book until the time of the end, when many will rush here and there, and knowledge will increase. (Daniel 12:4 NLT)

Your friends who simply use the *play* and *stop* buttons with their technology will miss huge opportunities to change the world. The increase of knowledge is so expansive that it requires the *pause* button. Imagine for a moment how you can multiply your opportunities through technology. What is God up to right now as He works in the messy areas of your life? What small changes are occurring? How can technology assist you in spreading His help to others?

Toward the end of the huge family Christmas gathering, when the last of the gifts were being opened, Lauren snuck into the closet in her bedroom. She hauled out a bag containing sixteen identically wrapped gifts. Trying to contain her smile, she began to distribute them to her extended family members.

"Better open them at the same time," she said.

Her mom gave her a puzzled look, then began to unwrap the book. The others followed suit.

They held a customized book, fresh from an Amazon printing company, with an extraordinary collection of old family pictures and a family tree. Page after page, pics overlaying pics, descriptions written in the margin explaining the who, what and where.

A hush fell over the room and she watched her grandmother wipe away a tear. In a family where it was hard to give gifts for the simple reason that most people had far more than they needed, Lauren used technology to give them a life gift.

●●●●●●●●●●●●●●●●●●●●●●●●●●●●●●●●

POSSIBILITIES:

- Which of the following areas could you see yourself exploring as a way to expand your influence through technology?

- ❶ Make a short YouTube video of your story and how you're learning to love Jesus.
- ❶ Pick one social media (Twitter, Facebook, Instagram) for advancing good stories about God and others.
- ❶ Choose one victim of cyber bullying per day and go against the flow. Make encouraging comments through social media about them.
- ❶ Self-publish a book.
- ❶ Text an encouraging verse to a hurting friend.
- ❶ Blog some of your journey with God and His truth.

– Choose one area in your life where technology is expanding in the wrong direction and replace it with a dream of the right direction. Write three small goals for yourself here:

1.

2.

3.

05/

Girl Power – Using What You've Got To Get What You Want

Liana went through her closet one more time, hanger by hanger. Big day at school. Lots of attention coming her way.

Yeah right, she laughed, *that dress would do it*.

She skipped over it and finally settled on one that was kind of in the middle.

Everyone knows the truth. Everyone agrees. A girl's body is powerful. What people disagree on is how she's supposed to use it.

Technology has not only magnified the issue, its magnification is something like 10 to the millionth power. And there's the battle you face. Two very different lines of thought are streaming from our world of technology and they only agree on one thing: *the extraordinary power of a girl's body*. Then comes the disagreement:

Strategy #1 – The girl *gives her body* to get what she wants.

Strategy #2 – The girl *guards her body* to get what she wants.

Pause.

Two opposite strategies, two radically different outcomes.

Try to guess the year that George Gilder wrote the following:

The prime fact of life is the sexual superiority of women. Women transform lust into love; channel male wanderlust into jobs, homes and families; change hunters into fathers; divert male will-to-power into a drive to create.

Eighteen Ninety-Six

1896.

That's a long time ago.

Want a translation? Try this...

A girl's body is powerful. If she values and protects it, it will draw a guy into marriage and convert his lust to love. He will change from a promiscuous, self-centered time-waster into a productive member of society.

The giant disconnect today is the separation of a girl's body from her inner being. If you've ever seen what strip mining does to land, you'll get the picture. Giant machines dig up the earth, stripping it of the mineral they're in search of. But they leave a mess behind them.

That's our world.

The power of a girl's body has been mined and commercially exploited. It sells the product. Most advertisers use it. And the mess it's left behind resembles piles of discarded, unusable dirt:

- The pornography industry
- Cohabitation where men are unwilling to commit to marriage
- Hooking up
- Violence and abuse toward girls
- Aggressive girls

Technology is messing with sexual identity, with the deep parts of what it means to be feminine or masculine.

But not the smart girl. Instead of just using the *play* button and going along with the flow, she pauses, thinks and connects the dots so that she avoids ruined relationships. Check out this verse in the Bible about the inner life:

I pray that out of his glorious riches he may strengthen you with power through his Spirit in your inner being,... (Ephesians 3:16)

C.S. Lewis and a number of others are credited with this quote:

You don't have a soul. You are a soul. You have a body.

The point is that we're deep people, comprised of far more than just mere bodies.

I've been a youth pastor for twenty-five years and once in awhile, I see extraordinary things. On one particular trip, we were out of the country doing a mission project. During the day we'd build houses – hot, nasty, sweaty, buggy. But at night we'd relax – resort, pool, Caribbean.

I came around the corner one night and found three guys huddled up, talking low. Just above a whisper. I stepped into the group and discovered what was about to go down.

The lead guy – a strong influencing-type dude – was upset. Apparently one of the girls had walked up to him at the pool, got within a foot of his face and said, "Do you like my locket?"

Innocent question right?

Ummm, not so innocent.

She was wearing a hot bikini, made from a mere scrap of material. He was 6'1", she was 5'5". She knew exactly what she was doing and her locket hung right there. Right in the middle of canyon walls. In order to see the locket he had to look straight down.

So he decided he wasn't going to be used and teased and led around like a dumb animal all week. He created a plan, got two other guys and was right in the middle of it when I walked up.

They decided to ignore the girls who were using 'girl power' to manipulate them and pursue the more modestly dressed girls.

I almost don't believe the words I'm typing, but I have to. I witnessed it.

The entire female clothing thing changed within two days. Without a word being said. Swimsuits were more conservative; t-shirts and shorts went over swimsuits when card games were underway. Evening attire was more modest. All of it, without a word.

This guy? He *paused* and formed a plan without just using the *play* button.

I'm going to ask you a simple question and you're going to answer. Picture that girl in her bikini. She walks up to America and says, "Do you like my locket?" What does America do?

Strip-mining for profit. Use her, then discard her.

FIFTY SHADES

Fifty Shades of Grey is the fastest selling paperback in history. I'm tempted to tell you what button to hit but I think you already know.

When the body of a girl is strip-mined for pleasure and the rest is thrown away, it makes sense that our world is out of touch with a girl's soul. Insecurity is getting so huge that the average girl feels more and more invisible. Therefore, culture gets more bent. The surprising thing about Fifty Shades is not how wrong and dark it is, but that it's a runaway bestseller. Women are reading this thing to explore why they feel so missed and what they might or might not be willing to do in order to be noticed.

Check out this amazing quote by Dr. Larry Crabb in his book *Fully Alive*:

> *Inviting, and no one comes. Opened, and unentered. Nourishing, and people go elsewhere. Relational femininity is risky. A devastating conclusion seems unavoidable: relating as a feminine woman is a stupid attempt to reveal beauty that does not exist.* (pg. 92)

I first heard the phrase 'emotional pornography' from a high school girl. Mazzie sat on the thick carpet, near the fireplace. Small group was underway and a dozen high school students were sitting in our living room.

"I'm done with emotional pornography," she said, before everything went silent.

Pornography was kind of a shame word back then and it was obvious that no one expected something like that to come from someone like her.

"What do you mean?" I asked, finally speaking into the silence.

She looked around the group, "Just saying...it doesn't work for me. Fantasizing with pretend mind games is destroying me because I'm creating these unrealistic expectations of how guys should notice me. When they don't, I cycle down into a bad place."

Mazzie had *paused* in her life long enough to see how foolish it was to pursue a love that doesn't exist.

For many girls, the pretend world is a powerful drug that further disconnects them from reality. Take Pinterest, for example. A girl might use it for the virtual arrangement of life. "Just add husband." But it's still pretend.

Historically, women used the power of their body to convert a guy's lust to love. They understood their own sense of power. They were life-givers, generation creators, worldview shapers and power transmitters. They were in touch with their feminine power. After all, the deepest part of any woman is the ability to give life. To give birth. To shape, nurture and influence. It's what makes her unique. And it's her true power base. To strip-mine her, using only her body, is to turn her into a half-girl. Who wants that?

Before 1950, the price of a girl's body remained pretty constant. If you lived in 100 B.C. or 1727, it cost exactly the same. It cost the man a lifetime of commitment. Yep, everything he had, every dream, every adventure, everything. For him to get the body, he had to trade in everything for this thing called marriage. Not now. Men don't need to sign a contract. Commitment is boring. Because, for every three girls, two of them are giving their body away free. A bunch of half-girls walking around, separated from their deepest sense of identity.

So this idea about the extraordinary power a woman possesses is not a new thought. What is new is the separation of a girl's body from her own soul. And much of this social surgery is carried out through technology.

– *Pill* – Contraception, including its most popular form – the pill – is a fairly new technology. Got approved around 1960. You'll discover a couple of things about girls who use it to avoid pregnancy before they are married: 1) It destroys a woman's uniqueness in that she starts to behave like guys. She gets more aggressive with guys since there is no consequence to having sex. 2) She is increasingly viewed in terms of a body. This destroys the view that a girl is unique and valuable and instead, turns her into an object.

– *Porn* – Technology took the separated girl (body only) and began to add other elements of power: 1) the power of the immediate, 2) the power of privacy, 3) the power of dopamine, 4) the power of shame at the addictive level, 5) the power of separating the user from accountability.

One of the strangest ironies of all time is how the feminist movement approached this idea of women and power. In their view, masculinity was the huge problem. Men dominated women and were therefore predatory and demeaning. They dominated at home, at work, everywhere. What women needed, they thought, was to be liberated and rescued from this domination.

Fast forward to today.

The feminists have brought big-time change to our culture. Some of it good, most of it not so good. Through the assistance of technology, a girl can wield the power of her body and experience a type of domination in her own right. She can use it, tease with it, lead with it, get promoted with it and spread it around wherever she chooses.

Alpha girl.

Irony.

By developing her body's girl power, she's become separated from her true self. She traded in femininity for promiscuity. And in the end, she lost all her power.

Irony.

Teresa entered her apartment and heard the familiar noise of electronic gunfire. "I'm home," she yelled, waiting for a response. She stared at the dirty dishes before turning around and heading out to the mailbox. *Clean up, make dinner, call the mechanic*, she thought to herself, mentally recording her to-do list as she shuffled through the junk mail. The noise greeted her as she re-entered the small kitchen again. They'd been together for two months now, Billy working a part-time job at Jimmy John's while he looked for a real job. Teresa – two years younger than him at 25 – was the leading sales rep for an adhesives company and seemed to make money as easily as falling out of bed. "Billy, I'm home," she yelled, louder this time, trying to break into the noise of the video game.

America's *play* button looks something like this:

- ⓪ Get good grades.
- ⓪ Dress in a way that will make the most of your body power.
- ⓪ Use it to get the guy's attention.
- ⓪ Trade it away to jumpstart a relationship.
- ⓪ Use contraception because abstinence is outdated.
- ⓪ Put marriage on hold until you get through school.
- ⓪ Kids are a pain in the butt anyway, and very expensive.
- ⓪ Get a degree.
- ⓪ Choose a career.
- ⓪ Then, maybe later, you can do the family thing.

IN THE BEGINNING

The Bible captures the story of a fatal virus that infected humanity when we mounted our rebellion against God. Interestingly, that virus took on a different symptom in women and men.

Placed in their own garden paradise, Adam and Eve lived in perfect relationship with God. Then, at some point, Eve entered into a dialogue with an extraordinary angel disguised as a serpent (later to be known as the devil). Her husband stood nearby as the conversation entered new territory – the possibility of becoming like God. The stipulation? Eve would need to do something that God had clearly told them not to do.

When the woman saw that the fruit of the tree was good for food and pleasing to the eye, and also desirable for gaining wisdom, she took some and ate it. She also gave some to her husband, who was with her, and he ate it. (Genesis 3:6)

Both Eve and Adam made a choice, plunging the human race into a disconnect from God. That relationship was broken by sin, a virus that proved fatal.

Notice Adam. He remained next to Eve during the temptation, knew it was wrong and yet said nothing. He later blamed her when questioned by God. Both the female and male patterns of fallen humanity can be traced back to that decision:

Girls – Control Through Pressure
Guys – Avoid Responsibility & Shift Blame

The girl who 'gets it' is the one who is well put together. She understands the female tendency to arrange and to apply pressure. She gets technology – how it can speed up her desires. But she also knows that the inappropriate use of it is toxic.

So she combines her inner life with her body in such a way that she values the whole package. She is able to let the half-girls run ahead of her because she knows that they'll wear themselves out. The race is not a 100-meter sprint. It's a marathon. And she'll win.

The Christian fights against anything remotely resembling the half-girl. Our opportunity is both timely and potent. Consider the power of a life that is distinctively set apart from the half-girl world, where the whole person is developed and valued.

This is not complicated.

Pause.

God developed it and it works out just fine if we let it.

SOUL AND BODY

At some point between a girl's 15th and 30th year, her body will tell her something so all-consuming that she'll do anything to obey it. It'll whisper so loud that all other loves will be drowned out. For 9 out of 10 girls it will say, "The reason I exist is to give life."

It can be ignored. For a little while. But eventually...

Michaela came into my office because she 'needed to talk.' And without any introduction she started in with the heavy stuff. Tears ran from north to south before dropping on her jeans. She ignored them – ignored the Kleenex box sitting next to her on the table.

She would soon be turning thirty-three and it had suddenly become clear to her. Too clear. Although she was quite successful in her career, it no longer satisfied. She wanted a baby. And that required a few other things – things like a man and a wedding and a marriage. But her body clock was nearly out of time. Where did this come from? And how did it sneak up on her?

CAREER VS. CAREERISM

For a girl, a career is a good thing. Very good. The tricky part comes in how she puts things in order. If her career goals get put at the top of the priority list, it's called careerism. And once that dominates the list, she's like toast caught in the toaster.

The biggest lie ever to come down the railroad tracks is that a girl can do it all. She can put all the family stuff on hold until she establishes girl power in the corporate world. And once that is in place, she can always do the marriage and kid thing. But it doesn't work that way.

Women today are accelerating way beyond their male counterparts. For every two college degrees that men get, women get three. And women now hold a majority of managerial positions in the country. This isn't a good time to be a girl looking for a career. It's a great time! But there's a whisper of caution if you listen carefully.

Somewhere along the line, achievement can become a threat to a guy depending on how it's used.

Notice the priority difference below:

Option 1 – Career First, Family Second.
Option 2 – Family First, Career Second.

Study these before we dissect them. What emotions are you feeling? What arguments seem to be forming in your mind? Why? How has this become such a passionate subject?

Most people read the two options in sequence. What should I do first, what comes second? But we're talking priority and not sequence.

Lacy stepped through the security gate, waited for the guard to check her I.D., then walked slowly down the hall. The last time she'd been here – the Children's Hospital – she'd been a child in serious pain, for the better part of a month. Impacting her so deeply, she'd chosen a nursing career and had just been hired into her first real job. Lacy loved kids, loved bringing joy and hope to their lives, loved nursing.

Over the last few months she'd taken a couple of hits. Her parents seemed to be fighting more these days and she'd gotten into an accident and totaled her car. School – the one thing she wished would end – seemed to be harder near the end.

She shook it off, peeked into a room and quietly entered so as not to scare the kid.

"Hey Tommy," she said, glancing at his name on the white board, "how you feeling today?"

Lacy is a complete girl, pausing long enough to feel the pain of life, yet redirecting it to serve others. Her beauty flows seamlessly from her inner being through her body and out her eyes. The little boy Tommy, with a shaved head and moist eyes that are trying to be brave, locks onto her and by the time she leaves, is a tiny bit braver.

PURSUE THE GREATER DREAM

God invites you into a world of imagination – a real one. He sets the fences of this new world in place and wants you to dream, to imagine, to build it out in your mind. Hebrews chapter eleven is all about this life of faith – the ability to see what you can't see:

> *Now faith is confidence in what we hope for and assurance about what we do not see.* (Hebrews 11:1)

It's impossible to out-dream God when you work within His formula. It's all about *who He is* and *what He's doing*.

You know the things you think about when you're falling asleep or sitting in a boring class? Yep, those things. Who created your ability to imagine? Who created the longing inside you? Answer this question: What is more awesome – the ability to dream or the one who created that ability? God is at the end of all our imaginations, waiting to engulf them in the greater reality of His presence.

Go for it.

⦿⦿⦿⦿⦿⦿⦿⦿⦿⦿⦿⦿⦿⦿⦿⦿⦿⦿⦿⦿⦿⦿⦿⦿⦿⦿⦿⦿

POSSIBILITIES

– Guys, how do you guard yourself against the half-girl menus of:

- ⦿ Soft porn magazine covers in the checkout lane.
- ⦿ The mall's sophisticated package of the half-girl.
- ⦿ Images, images, images.

– Girls, how do you guard yourself against:

- ⦿ The emotional pretend world where I lie to myself about a 'love that doesn't exist'.
- ⦿ Erotic books that explore explicit content through private e-devices.

– Guys, think about viewing girls primarily as unique persons, complete with a set of vulnerabilities, concerns, dreams and a growing need to relate with God. How are you doing with:

- The girl running the cash register. Is she a person or an object? Are you more focused on her clothing and body or her value as a person?
- Classmates. Where are they struggling the most and how can your prayers be more specific?

 – When talking with friends about girls, how much do you discuss how they think, work and love versus their looks?

– Agree or disagree? When a career turns into careerism, a girl betrays her inner soul and sets herself up for a life of regret.

 – What would have to change in your life to make sure that you're at a much better place by this time next year? And do you want that change?

06/

Guys – Risk, Action & Competition

The commercial goes something like this. The guy is not paying attention to where he's driving and crashes into another car. The driver of the second car is a woman. Neither is seriously hurt. As they exit their mangled vehicles, the guy is obviously stupid. He's dressed like a bum and tripping over words. Doesn't know what to say, doesn't know what to do. The woman on the other hand is self-assured. Stays calm and immediately calls her insurance agent.

In the magic of commercials – they only have 30 seconds – the insurance agent appears out of thin air. She too is an amazingly well dressed person. Very well put together. The two women solve the problem – again, in under 30 seconds. As the commercial ends, the guy gets this stupid grin on his face because he's been rescued by the two women.

As you watch movies, TV and commercials, keep track of how men and women are being portrayed:

- Stupid guy, smart girl
- Sloppy guy, neat girl
- Unemployed guy, great-job girl
- Insecure guy, secure girl
- Poor guy, rich girl
- Uncertain guy, self-possessed girl

A guy was born for *action, competition and risk*. I'm not lying, unless you talk to a feminist. In that case, I'm lying through my teeth because she'd say that differences between boys and girls are purely a result of social influences when the children were tiny.

You know the story. Parents give their son a truck while their daughter gets a doll. The boy rams the truck into the wall while the girl cradles the doll and attempts to nurture the thing. If they'd reversed the gifts and if they'd kept the poor kids from being traumatized by outdated roles, the boy would cradle the doll and protect it while the girl would drive the truck off the back of the couch and burst out in a wicked laugh as it did three somersaults before coming to rest on the carpet.

Parents laugh at this fairy tale, this fabrication of the gender-equity movement. They might agree that their son wants a doll but for a very different reason. He needs a sword to bash things with and can't find his at the moment.

You could also ask a comedian, if you need to be absolutely sure that guys and girls are different. In his book, *Dave Barry Is Not Making This Up*, Barry writes:

> *I hate to engage in masculine and feminine stereotypes, but when women plan the menu for a recreational outing, they usually come up with a nutritionally balanced menu featuring all the major food groups, including the Sliced Carrots Group, the Pieces of Fruit Cut into Cubes Group, the Utensils Group, and the Plate Group. Whereas guys tend to*

focus on the Carbonated Malt Beverages Group and the Fatal Snacks Group. On this particular trip, our food supply consisted of about 14 bags of potato chips and one fast-food fried-chicken Giant Economy Tub o'Fat. Nobody brought, for example, napkins, the theory being that you could just wipe your hands on your stomach. Then you could burp. This is what guys on all-guy boats are doing while women are thinking about their relationships. (Pg 133)

Pause.

What is the gender-equity movement really trying to accomplish?

Manhood is under attack.

War.

Two sides are lined up in battle:

Historical	Feminist
Risk	Risk Averse
Competition	Competition Free
Action	Sit Still
Experience	Feeling

One of the coolest stories I've ever read is about a guy who got his doctorate degree in political philosophy and ended up with a job as an Executive Director of a Washington think tank. Interpretation? He went to a heck of a lot of college to get a job with a heck of a lot of talking to a bank where he cashed his weekly checks for a heck of a lot of money. Got it?

But it didn't last.

Not because he wasn't good at it but because the job sucked. The lifestyle sucked.

Want to know what he ended up doing?

Fixing motorcycles.

Listen to what he says:

The wad of cash in my pants feels different than the checks I cashed in my previous job...I was always tired, and honestly could not see the rationale for being paid at all – what tangible goods or useful services was I providing to anyone? This sense of uselessness was dispiriting. The pay was good, but it truly felt like compensation, and after five months, I quit to open the bike shop. It may be that I am just not well suited to office work. But in this respect I doubt there is anything unusual about me. I offer my own story here not because I think it is extraordinary, but rather because I suspect it is fairly common. I want to do justice to intuitions that many people have, but which enjoy little public credit. This book grows out of an attempt to understand the greater sense of agency and competence I have always felt doing manual work, compared to other jobs that were officially recognized as "knowledge work." Perhaps most surprisingly, I often find manual work more engaging intellectually. (Pg 5, Shop Class as Soul Craft)

By working with his hands, he actually thinks better about life and feels better about his contribution to it.

There is something amazing going on inside your brain. It's called blood flow. In a natural world, here's how it works. When you try something risky or something that requires hard work, the blood flow divides nicely. Part of it goes to the risky side of the brain and part of it goes to the reward side of the brain. Risk and Reward. Balanced. Kind of like a good partnership.

Here's what's not amazing. Technology has figured out a way to steer the blood to the reward side of the brain without the necessary risk part of a boy becoming a man.

Joe looked at the river again and then at his two friends who'd already crossed and were yelling back at him. His legs felt like overcooked macaroni and a sense of dread settled over him. The water raged and looked like chocolate milk from a recent storm but they'd somehow managed to get across.

Suspend this scene for a moment.

What's going on in Joe's brain in terms of blood flow?

Lots.

As he steps into the water, stumbles a bit and gradually figures out how to brace his legs against the torrent, a confidence begins to build. At some point he crosses the tipping point to where he thinks to himself, "I can do this."

When his right foot finally touches the other bank, suspend the action again and look inside his brain. The blood, which was flowing into the risk section, has now diverted and is flowing through the reward section. Pleasure. Pride. A huge rush.

That's real life.

But technology is offering him the same pleasure without the experience and risk of crossing the river.

VIDEO GAMES

In an extraordinary breakthrough, game developers know how to manipulate your body through video games. They can literally trigger your body to release dopamine at strategic points in the game. By advancing through these games, you can get a real world sense of rush and pleasure without the real world risk.

So what. What's the big deal?

So what?

Here's the so what. Feed yourself a steady dose of reward, without risk, and you'll become a junkie. Like any drug, it steals your life, burns out your reward center and leaves you depressed. Think about the drug addict. There was no food in the refrigerator because he was a moron and didn't work. So he got high. When he came back down, someone had stolen the refrigerator. That's the so what.

Kevin raised his hand, which I thought was pretty unique since there were a couple hundred adults in the auditorium and...17-year-old Kevin.

We'd been discussing video games – the good and the not-so-good things about them. That's when Kevin decided to share his opinion.

"I'm a gamer," he said, as many turned to listen. "I grew up in the video game culture and could play all day, without coming up for food."

I interrupted, "And you're going to med school so that you can operate on me in my old age?"

This was actually true. Most of the people in the audience knew Kevin to be extremely bright and in the early stage of researching med schools. That's why we were eager to see where he was going with this.

"It's true," he continued. "But part of the process of discovering what I wanted to do came through video games. You see, I'm good with my hands and I love solving problems. I also love serving people. But I had to start telling myself when to say when. I could actually tell that I was getting a cheap reward for doing nothing."

He sat back, with several in the audience nodding. I don't know...maybe they were mentally calculating whether or not they wanted to go under anesthesia with Kevin holding the knife. But he made a good point.

How do we learn to navigate a technology world where we can get free hits of dopamine? That's the big question. Without working through this challenge, it leaves us open for bigger, more powerful hits of dopamine and an eventual depressing life that disconnects risk and reward.

PORNOGRAPHY

Visual porn is like video games on steroids. More power, more pleasure. The real world of guys and girls goes something like this. Boy likes girl so he figures out how to move towards her. He wants to figure out if she'll like him in return and it causes him great concern. Lots of blood pumping through the brain due to the huge risk required. The more he likes her, the greater the risk appears to be.

She may reject him and the very thought stops him dead in his tracks. It's only after taking the risk that he discovers one of two things: 1) She likes me. I pulled it off and she likes me! 2) She rejected me but I'm still alive. I survived and I can handle this as I figure out my next approach.

In the real world he wins either way.

Blood flows between risk and reward.

Not true with porn.

There is no risk.

It's all reward.

In the fantasy world of pornography, the girl always wants you. You pretend like it's a hunt. Like it's a risk. You're searching, you're clicking, you're hunting for a girl. But that's bull crap and deep down you know it. There is no risk because you're playing a game where the girl always wants you. It feels good. There is momentary pleasure. Then the game ends.

Each session leaves you weaker in the real world, less able to move toward real girls, more depressed. In that state of mind it's easy to feel like you need a greater hit of pleasure, a more potent drug, a deeper porn session. That's why sin is genius. It creates this downward cycle of self-destruction and doesn't even need to stop for gas.

Here's something that should get your attention. All our technologies trigger dopamine hits of pleasure. In fact, guys' brains are being digitally rewired towards adrenalin and change and excitement. Think through the following options. In what way does each offer a greater reward than risk?

- Surfing Craigslist
- Texting
- Twitter
- Watching a movie
- Porn
- Fast food
- Pot
- Video Games
- Facebook
- Buying something on credit

Whether it's something you put into your mind through the eyes, or something you put into your body through the mouth, you'll notice that our culture is scientifically separating risk and reward and stuffing us full of rewards. Pleasure. Temporary.

Why should you care?

Because real relationships with God and others develop subtly and slowly.

RISK

The real world involves a balance between risk and reward. Oh, that's right, I've already said that three times. You have a good memory. Girls have built-in risks, but guys have to create them. A girl's body forces her to go through painful things (monthly period, pregnancy, childbirth) and she learns to grow up. Not true for a guy. He needs outside forces to create the same life lessons.

Until very recently, most civilizations forced their boys into male initiation rites. They knew a simple truth about life: Men are not born, they're made. Richard Rohr says that cultures made boys go through all sorts of crazy stuff until every boy could say this about himself:

- Life is hard.
- Life is not about me.
- I'm not that important.
- I'm not in control.
- I'm going to die.

The guy who just uses the *play* button might glance at that list and think it's negative. It's not until he *pauses* and reflects on it that he begins to see underneath and around the words. This list is the reality of our world – a broken place where hard things exist and we're not in control. The list is ultimately freeing because it draws a guy's hope to God.

Aaron experienced the type of childhood that no one should ever have to go through. By the time he was thirteen, he'd witnessed a gunfight, a lawsuit that tore his family up and the death of a sibling.

During his third year in college Aaron began to seriously hit the *pause* button. His life was a mess, his relationships toxic, his finances in the toilet. He needed help and began to reach out to some older guys, people who'd gone through some deep stuff and had lived to tell about it. In fact, they were living well.

Aaron's life slowly began to turn. Inconsistent at first, up and down, hypocritical. He put God first and began to experience that strange sensation of moving parts – some very unpleasant – being directed by someone other than himself.

It's three years later and he's now doing the same thing for younger guys. They look up to him because his relationships are stabilizing and he uses his extra money to surprise people with help from time to time. The ones behind him on the trail notice stuff like that. I asked him the other day what his favorite verse in the Bible is:

Set your minds on things above, not on earthly things. For you died, and your life is now hidden with Christ in God. (Colossians 3:2-3)

Cliff Graham, author & screenwriter, posted this on Facebook the other day:
Being a man is supposed to be difficult.
The title is earned, not biologically developed.
It implies selflessness and sacrifice and should leave us utterly exhausted if we're doing it right.
Throw your shield down in front of the weak.
Love with great ferocity.
Give grace generously.
Attack evil relentlessly.

●●●●●●●●●●●●●●●●●●●●●●●●●●●●●●●●●●●●●●

POSSIBILITIES

Which of the following have you done? What did you learn from the experience? Which item on the list could turn into your next risk?

- Keep something alive for a year.
- Impose a self-limit on the time you invest in video games.
- Take a six-week trip out of the country.
- Save up and buy something with money you earned.
- Talk to a girl who is driving you crazy and communicate that you're not interested. But here's the risk: 1) face-to-face and 2) do it in a way that builds her up.
- Talk to a girl you're interested in.
- Ask a girl out.
- Pay for college without borrowing money.
- Take a machine apart and put it back together without any leftover parts.

- Read a book that isn't part of a school assignment.
- Find your way around a big city alone.
- Plan a road trip with friends and do it.
- Build a campfire without gasoline.
- Sleep outside in the winter.
- Conquer a 14,000' mountain.
- Shoot a gun.

07/

out-DATED

Dating is outdated.

No it isn't.

Yes, it is!!

I know that some people still date. I know that people who end up married started the relationship by dating. I know. But girls, you need to stop measuring yourself by an outdated system. Stop putting the beat-down on yourself because you don't get enough dates.

In 1811 and 1812 there were a number of earthquakes around New Madrid, Missouri. The ground underneath the Mississippi River rose so significantly that the river actually started flowing backwards.

That's the deal with marriage today. People have lost confidence in it and it's out of favor and it's flowing backwards. So what does that mean for you?

Dating is only as strong as the object it leads to. When we date, we're playing *dress-up* for marriage. From 1950-2000, the dating river flowed towards marriage. Almost everyone jumped in and rode the river to marriage. But the river started flowing backwards, dating began to be outdated, like water that wasn't flowing anywhere.

– *Suzanne* – She measures her desirability by how many times she's getting asked out. But other than homecoming and one other *sort-of* date, she's not getting much action. Instead, she hangs out with groups. Dating as a measuring system is outdated. Her beauty and amazing personality can no longer be held hostage by whether or not someone asks her out.

– *Paul* – Social networking allows him to eliminate risk-taking with girls. He likes a particular girl but finds it easier to hang out online or in groups. Dating as 'risk taking' is outdated.

– *Mrs. Carpenter* – She's privately relieved that her daughter isn't dating because of the messy things that come along with it. But she's only focused on now. She hasn't really thought through the issue of the decline of dating for when her daughter is twenty-five or thirty. Her view of dating is outdated.

– *Mr. Jackson* – He's pretty conservative and really doesn't want his son doing much dating until college. His belief is that it'll preserve his son's purity. But technology delivers infinite opportunities to trash one's purity. More than dating ever did. In this way, his system of protecting his son is outdated.

In a powerful way, dating was built on the confidence in marriage. Yes, yes, yes...dating wasn't perfect. We all know that. But it did move people towards something they desired. Marriage.

The decline of dating is another indicator – like a flashing red light – that marriage and childrearing are out of favor.

NO RISK, NO CONFIDENCE

Seth was fifty yards away but I could tell he'd spotted me. He was working his way across lifelines like a man on a mission. There were twenty sailboats rafted together in Thomas Cove with two hundred students and leaders aboard. We were on a weeklong sailing trip in Canada's North Channel.

"Hey," I said. "What's up?"

He looked over his shoulder and waited until he'd closed the final gap, "Can we talk?"

I nodded, so he proceeded. "I was wondering if we could do dialogue-dating tonight?"

I heard his words but his eyes said it all. He was interested in someone and wanted a chance to talk with her. Our students enjoy *dialogue-dating* – a structured time that we create for them. We give them a question and have them meet with someone of the opposite sex to talk. Every twenty minutes they switch dialogue partners until the time is up. But, since this wasn't in the schedule for tonight, my brilliant mind came up with an extraordinary solution.

"Why don't you go talk with her now?"

Poof. His smile was gone.

He studied his shoes. "Um, that's okay. I don't want to send the wrong message."

Seth is not unique. He's a guy who is interested in a girl, but rather than pursue her – which would require risk – he's looking for a third party to make it happen.

You should probably know that we've been doing trips like this for about twenty years. Ask your parents what dating used to be like. Did they go to summer camp or do a youth group trip? In the old days, trips produced lots of dating relationships.

Not now.

I often see several hundred students hang out for a week without one couple forming.

Right or wrong?

No.

Different?

Yes.

Mark Regnerus, an associate professor of sociology at the University of Texas-Austin, is co-author of *Premarital Sex in America: How Young Americans Meet, Mate, and Think About Marrying* (Jan 10, 2011). The book is based on extensive data, collected from four national studies representing a total of 25,000 young people, ages 18-23, and more than 200 additional interviews. His study of the college campus environment reveals a higher percentage of women to be a leading reason for the increase in hooking up.

The women wind up competing with each other for access to the men, and often, that means relationships become sexual quicker...Men don't have to work as hard as they used to, to woo a woman...I've talked to various interviewees who had never been on a date, which doesn't really make sense, given they're pretty attractive. It's just that less seems to be required to be in the company of a woman.

So what happens to the young people who are caught in this revolution? Where are they heading and how do they get there? According to Leon Kaas, in his extraordinary article *The End of Courtship*,

Today, there are no socially prescribed forms of conduct that help guide young men and women in the direction of matrimony...People still get married – though later, less frequently, more hesitantly, and, by and large, less successfully. For the great majority, the way to the altar is uncharted territory: It's every couple on its own bottom, without a compass, often without a goal. Those who reach the altar seem to have stumbled upon it by accident...

Scared yet? Check out his comments on college students:

Never mind wooing. Today's collegians do not even make dates or other forward-looking commitments to see one another. In this, as in so many other ways, they reveal their blindness to the meaning of the passing of time. Those very few who couple off seriously and get married upon graduation as we, their parents once did, are looked upon as freaks. After college, the scene is even more remarkable and bizarre: single bars, personal "partner wanted" ads, men practicing serial monogamy (or what someone has aptly renamed "rotating polygamy"), women chronically disappointed in the failure of men "to commit."

Here's the last one. I promise. Dr. Paula England (Professor of Sociology, Stanford University) & Dr. Rueben J. Thomas (Asst. Professor of Sociology, City College of New York) write,

Dates are no longer that common and people mostly hang out with friends or hook up. Today, students on college campuses state that the traditional date is nearly dead.

Okay. You get it, I get it. Not a lot of dating anymore. But the important thing is to recognize the role that technology plays in the shift. Those who allow technology to manage their intimate desires are becoming disconnected at important levels. They attempt to gain the benefits of marriage (sex & intimacy) without all the work and commitment. They are strip-mining sex and discarding marriage. A serious disconnect.

Relationships *are* risky but relational connection must follow the code. Like DNA, God stamped our soul with the following:

Boys to Men – This is a process that requires hard work. Boys are not born, they're made. Through risk-taking and action.

WARNING: The cultural story of girls giving their bodies to boys ahead of time imprisons boys in a perpetual state of boyhood. They don't grow into men.

Girls to Women – This is a process whereby girls learn to tame their desire to control and direct relationships, creating a power that causes boys to risk.

WARNING: The cultural story of boys turning girls into pictures and images and objects of pleasure imprisons girls in a perpetual state of half-girls. They don't grow into women.

Both guys and girls are equally tempted to take the easy way out. Which of course, isn't easier at all when you consider the long run. Let's revisit the two types of pornography that dominate the virtual landscape. Which are you more prone to?

Visual Pornography (pV) – Creating a pretend world by combining pics and videos and telling the characters what to do for your own satisfaction. Except for a few small details: It avoids risk and traps the user in a pretend world.

Emotional Pornography (pE) – Creating a pretend world by reading and/or imagining a fantasy situation where you are the main character and others get told what to do to satisfy you. Except for a few small details: It avoids risk and traps the user in a pretend world.

Technology plays a major role in delivering both visual and emotional pornography. It's the carrier that brings the pretend world to you. In the misuse of technology we see the disconnect. But for those who *pause* long enough to get it, technology can be a great tool as they take risks in the world of guy-girl relationships.

One of my favorite verses which helps me evaluate and navigate the guy-girl world is:

*But among you there must not be even a **hint** of sexual immorality, or of any kind of impurity...* (Ephesians 5:3)

RISKS & TECHNOLOGY

Brad looked up from his plate and stopped chewing. His group leader had just challenged him in a way he least expected. He forced himself to finish his oversized bite of hamburger and asked, "What do you mean?"

Andrew, the leader, was staring out the window thinking, then turned back to him. "Do you have a girl who is driving you crazy? Texting you all the time and trying to get your attention?"

Brad rolled his eyes.

"That's what I'm talking about," said his leader, for the second time. "Before we meet next week, I want you to talk with her. Face-to-face. I want you to figure out how to tell her to back off. But here's the deal. You have to do it in a way that honors her."

Brad took another bite.

Brad is going to have to learn to deal with two types of girls. And each involves a certain type of risk. The question for Brad is deeper than dating and deeper than social networking. It's about becoming a man.

One type of girl, mentioned above, is so aggressive that she has ruined any chance of him liking her. In fact, it's creating the opposite effect. The second is the type of girl he's *very* interested in. In fact, she's so awesome that he can't even imagine talking to her without choking.

Enter technology.

He has to figure out how to use it appropriately without using it to avoid risk. While he could block the aggressive girl's cell number, there's a better way. Harder? Yes, but better. It honors her and helps her see what's really going on.

As for the girl he likes, he could text her. But again, that involves using technology as the easy way out. Is it wrong for a guy to text a girl? Nope. Is it wrong for him to avoid risk-taking by texting? Duh.

Her name is Brittany, she's in seventh grade, and she's good at what she does. She's an arranger. A social architect. An artist. Her social antenna is finely tuned and, like a pro fisherman trolling with a fish-finder, she intuitively locates groups of guys and manipulates through subtle pressure. Why? Because she's convinced that nothing will happen in the guy world unless she makes it happen.

If Brittany's trolling continues unhindered, her hook-ups will mature from the silly stuff of middle-school to the destructive variety. The virgin card, which girls used to keep locked away, somehow got reversed. In her confused state, alpha girl now plays that very card to get what she thinks she wants. And she plays it early. Our world is full of sexually confused, needy girls. They themselves are full of pressure from our consumer world of strip-mined images. They have no understanding of their core identity and dignity, their worth and value. These girls have not learned to develop their inner beauty in step with their outward beauty.

Over the years, different systems have delivered children to adulthood, young people to marriage. Arranged marriage. Courtship. Dating. Social networking. Sometimes it's one of these alone but usually it's a combination of two or more. It's how people meet, greet and move into relationships.

Your job is to avoid the disconnect brought on by excessive dependence on technology and embrace the risk-taking associated with God's stamp on your life.

OLDER & WISER

I haven't met one student in the last 25 years who thinks God is calling them to be single. So, we have two things going on at the same time: 1) Almost everyone hopes to get married and have a family someday, and 2) It's getting harder and harder to figure out how it's going to happen.

Here's what you can do about it. You can find a person who is older than you and wiser than you. Someone you trust enough to begin trusting them more.

Some of you have awesome parents while others experience lots of power struggles with mom, dad or both. A few of you have parents who are actually behind you in terms of maturity, parents who've messed up big-time.

The best-case scenario is to have trustworthy parents and then add another person, outside the family to your advisory team.

At age 13, Ben was lucky. Without thinking much about it, he had this guy, Tiny (6'4", 250 lbs), at church who was ten years older than him. Tiny always went out of his way to ask Ben how life was going. When he was younger, Ben wanted a phone but his mom kept saying, "No, not yet." So they'd talk about it. Tiny would give him some ideas and Ben would try one, if he remembered.

When Ben was sixteen he couldn't figure out how to live with girls and how to live without them. His world had grown incredibly complicated due to technology.

They talked. Ben listened some more because Tiny seemed to understand how he was stuck. He got some ideas on drawing 'boundaries' – a word he'd never thought much about before.

When nineteen and home from college on Thanksgiving break, they hit Starbucks and laughed for twenty straight minutes about all the dumb stuff they'd talked about years ago.

Then, they talked. Really talked.

Here's the bottom line. For thousands of years, cultures have had systems in place for boys to turn to men and for girls to turn to women. Dating was one of those systems, but it isn't enough now. You need a private _____. Call that person anything you want: superhero, coach, tutor, mentor, advisor, arranger, confidant, counselor, discipler, influencer.

You're creating a customized relationship.

How?

Pick one of your favorite people who is older and more experienced than you. Choose someone who's been through the kind of struggle that you need help with.

◍◍◍◍◍◍◍◍◍◍◍◍◍◍◍◍◍◍◍◍◍◍◍◍◍◍◍◍◍◍◍

POSSIBILITIES

- List three possible people who could potentially be your private advisor:
1.
2.
3.

- Describe the way in which your guy-girl life has become complicated. Where do you need the most help?

- What verse in the Bible speaks to your core struggle in the guy-girl world?

- Identify the exact place where you're tempted to use technology to avoid risks when it comes to this guy-girl area.

– Choose one risk that you'll take in the next seven days – a risk large enough that you need God's help to pull it off. Write it down, date it, and have a good friend or mentor hold you accountable.

08/

Cyber-Relating

John reread the comments under the picture the girl had posted because he wondered if he'd missed something. But no, it was all there, with seventeen people trying to outdo themselves. Each comment got worse.

He posted a quick, reverse-type comment but didn't really think it'd help much. Then he thought about texting her. He hesitated, not knowing what to do because he didn't want to open the door in case she blew up his phone with ten texts per hour. She could be like that. But still, she didn't deserve what they were saying about her.

He finally decided to do it. Grabbing his phone, he texted, "Hey...ignore it...not true."

People are built for relationship. But "relating" on the internet is rarely good "relating." Have you noticed that? The stuff of social networking usually involves one or more of the following:

Hider – I use the internet to avoid real relationships.

Chameleon – I change my personality depending on who I'm talking to.

Superhero – I portray myself in a superhuman way with subtle lies.

Stalker – I research people on-line.

Cautious – I communicate online to guarantee the perfect response.

Revenge – I transfer anger from life and past experiences to people online.

Numb – I cover loneliness.

Impress – I find the best stuff out there and re-post it, to make myself look good.

This list could be a mile long and we still wouldn't run out of ways that we fake-relate. It makes sense in a way because the Internet is still very young and we're still trying to figure out how to relate through it.

This is where you come into the picture. You're part of the generation that has an opportunity to humanize the whole thing, capturing all the positive aspects of it so that you can deepen relationships.

Using the situations in the left column below, try to match them with an appropriate technology in the right column. How would you go about each assignment?

A teacher gave you an unfair grade   Individual Text
Someone Facebooked a lie about you   Group Text
A friend tells you he's gay   Private Message
It's your dad's birthday   Instagram
A friend wasn't in school today   Facebook Post
You want to invite someone to church   Twitter
A friend's parents are getting divorced   Phone Call

You found a new coffee drink	YouTube Video
Planning a party at your house	Face-to-face

Our lives are all about relating, but digital interaction can reduce our relationships to the lowest common denominator – *information sharing*. It's easy, it's fun and it fires off some dopamine. Scrolling through Twitter feeds is like a massive information dump for me. Nobody can afford to waste words because there are only 140 characters allowed, right? So it gives a false sense that this is life, this is where the most important people hang out, this is where I should be.

Does the world really need another *selfie*? If you want to laugh, look up "instagram-quiet-time" on *stuffchristianslike.net*.

In a funny way, the video tries to portray the trending view of culture – things really don't happen unless we document them. And documenting them is more valued than truly experiencing them. On the short video, this girl attempts to impress the whole world by recording her quiet time with God and as a result, loses both. She has no quiet time and she's not been with God.

Big question: How do we use our technology to improve the way we were meant to relate with each other?

You've probably heard enough lectures about cyber bullying to last your whole life. Me too. Then why do we rip people off by limiting our interaction to information sharing instead of a meaningful exchange? I often pass the following verse through my mind before meeting with people. Sometimes I need to push it through several times – it takes me that long to get it:

> *May the words of my mouth and this meditation of my heart be pleasing in your sight, LORD, my Rock and my Redeemer.* (Psalm 19:14)

During his 17th year, Peter spent five weeks in a village outside Guadalajara, Mexico at a children's shelter. His job consisted of caring for cows and goats, shelling corn, doing dishes, playing with children, assisting in teaching and doing whatever the leader needed help with. His friends remained in Michigan.

Some of his relating seemed like an information dump. His family and close friends wanted to know what was going down. When the Internet worked, he reported facts. But there were other things transpiring, issues that demanded good relating. He was rushed to the hospital once and needed some real-time connection with his parents. Life on the compound proved difficult with different cultural issues and he needed to vent, to listen, to learn, to practice. All of this stuff is called relating and it consumes huge portions of our lives.

Technology offers a multi-layered benefit package when it comes to connection with people but it demands that we *pause*, think and change some of the ways we're using it.

Peter's blog dumped information to his family and friends, but his private email allowed him to *merge* with one person at a time. He wrote Cari and she wrote back as they identified similar feelings of loneliness and explored ways to deal with it. That's relating.

Cari *re-tweeted* one of Pete's problems, asking her friends to pray for him. Even though this is a form of passing information along, it also takes advantage of technology by building a network of people who care. Before she shared this information, she got his permission, requiring him to *cooperate*.

Notice the words in italics: merge, re-tweet, cooperate. That's what happens when you take information to the next level and begin to humanize it.

❶❶❶❶❶❶❶❶❶❶❶❶❶❶❶❶❶❶❶❶❶❶❶❶❶❶❶❶❶

POSSIBILITIES

– Try some of these ideas out on your friends sometime. See if they bring about a deeper sense of connection.

❶ Make a Restatement – If I heard you right, you're saying...
❶ Use a Follow-Up Question – Really? Then let me ask you this...
❶ Open Ended Questions – Make sure your question begins with *what* or *why* or *how*.

– Post one comment a day to create a reverse-trend wherever you see cyber-bullying.

– Find a person who's a bit younger than you and try to draw out some deeper conversation with them through their technology of choice.

– Find one person who's ahead of you on the path and ask them for advice on one of your current struggles.

09/

Same-Sex Attraction

Rina finally came out and, within a week, pretty much everyone knew about her same-sex attraction. Within a month, her inner circle of friends completely changed. Her best friend from childhood, Sara, eventually got Rina to agree to talk about the issue over coffee.

Sara watched the steam rise before attempting a sip, "What's happened between us?"

"A lot."

"Talk to me," said Sara.

Rina stared sideways, watching a car enter the drive-thru. "You're all judging me and I'm sick of it."

Sara waited until Rina turned back and looked at her. "How have I judged you?"

"Well," said Rina, "maybe not you so much as the others."

More than most issues, the challenge of same-sex activity and how to think about it requires a serious *pause*. The movement is so large and so well-funded, it requires that we step outside the argument to see what's going on.

Some judge, some don't. Some people err to the side of truth while others just try to love. But the battleground is so entrenched that sides form immediately. And that's what happened between Rina and Sara. One moment they were best friends, with a gazillion shared experiences from childhood. The next moment, without even talking about it, they became strangers.

Have you ever seen a great *tug-o-war* match in person? The kind where the losing team gets dragged through the mud-hole in the middle? You know how it goes. Team one lines up on one side of the mud, puts their heaviest people at the end of the rope, makes them get down low on the ground, wrap their arms around it and dig their heels in. "No way," they think, "is that rope ever going to drag our big guys the wrong direction." But team two does the same.

Opposite directions.

The gay-straight debate is a giant tug-o-war.

If Sara and Rina are going to rebuild their friendship, it will take a lot of work because it involves moving through conflict and living within the tension of truth and love.

Drew's reputation around school began to reverse itself. It all started when he got into an argument in class. With the teacher. Somehow, someway, the pre-calculus teacher found a way to bring up the same-sex topic every week. Seriously? In pre-calc?

People knew Drew was a Christian for a very simple reason. He told them. But an identity began to form as he argued with the teacher, and it didn't take long for the reputation to develop. A few of the arguments got heated and became counterproductive. This led to before and after school discussions where both he and the teacher involved other students. Sides. Teams. Tug-o-war.

It's a strange development. Christians, who are supposed to be identified by an extraordinary love, are now labeled as haters and as a gay-bashing group. On the other hand, gays and lesbians, many of whom have religious backgrounds, are now presumed to be on their way to hell because they're embracing their sexual orientation.

A giant mess.

Neither side wants to budge or yield an inch.

Team Love – These people are interested in listening to each other. Each story, each background, each person is important. Having a different sexual orientation is okay, even celebrated.

Team Truth – These people focus on what is true, as in, what God has to say about the matter. Feelings can't be put ahead of what's right and true.

You see the problem? Not only are they moving in opposite directions, but they begin with the assumption that the other team misunderstands them and has the wrong starting point. There are lots of words but no communication.

Want to know how two extreme sides form? It's called *avoiding the tension*. If the love team wants to enter the tension, they must be willing to look at how the truth might call them towards change. If the truth team is willing to enter the tension, they face the real possibility that the people they love may never change. You see, real love isn't based on whether or not the other person changes. Love is something you can't earn. God loves us from what He finds in Himself, not what He finds in us. If I decide to move towards a person – only long enough to figure out if they'll change – I haven't loved that person at all.

THE PROBLEMS WITH EXTREMES

To love *only*, is a shallow thing and in the end, isn't very loving.

- When we think it's loving to avoid telling someone they're wrong, we're just buying into the mindset of our culture.
- We don't do this in other areas of life (like when a child walks towards the road or when a gang wants to kill).
- To love another person is to want their life to go well for them.

To focus on truth *only*, is a terrible thing and, in the end, isn't very true to the other person. What is your source of truth?

- **❶** Wiki – Truth by consensus (when the most people agree, it must be true).
- **❶** Google – Truth by relevance (when it rises to the top of a Google search, it must be true).
- **❶** Bible – Truth by authority (when God says it, it must be true).

The Bible talks often about the balance of truth and love. Check out this verse,

> *Then we will no longer be infants, tossed back and forth by the waves, and blown here and there by every wind of teaching and by the cunning and craftiness of people in their deceitful scheming. Instead, speaking the truth in love, we will grow to become in every respect the mature body of him who is the head, that is, Christ.* (Ephesians 4:14-15)

Speaking the truth in love.
That's the goal.
How do we get there?

WHERE DO YOU STAND?

The following two quotes come from famous Christians. Applying these two quotes to the same-sex-attraction movement, which quote seems better to you?

> *It is the Holy Spirit's job to convict, God's job to judge and my job to love.* – Billy Graham

This triangle of truisms, of father, mother and child,
cannot be destroyed; it can only destroy those
civilizations which disregard it. – G.K. Chesterton

Some of us immediately lean towards one or the other. But some of us like them both. Read them again and place yourself somewhere on the following scale:

1	2	3	4	5	6	7	8	9	10
Love				Balance					Truth

Here's what the Bible has to say about sexual sin – and it might surprise you: The Bible states that none of us get our sexuality right. None.

The major section of the Bible that talks about our deviant sexuality is Romans 1:18-32.

God is angry. But instead of it being a temper tantrum, where He suddenly gets upset over something we did, it's a long-ago story. And here's where we get it wrong. We think that because we do the things on the list, He's ticked at us.

Nope.

Here's what it says, and to correctly understand sexual sin, you have to get the order right:

- **❶** God made everything.
- **❶** The average person – even using just a fraction of their brain – gets it. They look around at this amazing world and instantly know two things: 1) somebody is powerful and 2) somebody is awesome.
- **❶** But people were not interested in admitting that, because if somebody else is in charge, then it limits freedom. It means someone else gets to be the boss, and we don't like that.
- **❶** So people tried to stamp out the knowledge of God, to get rid of it once and for all.
- **❶** God got angry.
- **❶** God turned us over to run our own lives.
- **❶** We began to self-destruct.

Notice. We haven't talked about sexual sin yet. Where does that come into the equation? Well, once God got angry, He cut us loose. Basically He said, "You want it? You got it!" So this is very, very important. Sexual sin didn't make God mad. It's just something people started to do once God cut them loose. And once you see the order, you'll begin to see the downhill progression.

Not only did people begin to do 'such and such,' but they even began to do 'such and such.'

It just keeps spinning out of control.

Going downhill.

Remember this. Once God cut us loose, we began to trash our lives.

You're going to have all kinds of sexual urges, your whole life. You were designed as a sexual person. More importantly, the power plant that God put into your deep place – to run the entire operation – only magnifies sexual desires.

I remember being sexually ignited once when watching a scene on TV. Two girls, not overly dressed, began kissing passionately. Does that make me gay? Should I start questioning and exploring my sexual orientation? We all know that's what our culture would suggest.

Anthony got asked out last year. By another guy. The experience jarred him so much that he began to explore his sexual identity online. That's where he did the research and that's where he educated himself.

What do you think he heard out there?

What did the world tell him?

If we allow our urges and unique feelings to run the show, we begin the downhill slide.

Michaela is attracted to other girls. The feelings are so strong that she starts to feed them. They become stronger. It seems unfair for her to imagine any other way of living so she allows the lesbian label to become her identity. Across town, Jay is attracted to girls but in real life they threaten him. So over the course of two months, he becomes addicted to pornography. This label begins to define him. The world will tell you it's all okay and you can't judge anyone over issues of identity.

A lot of labels.

The same human condition.

Moral perversion is the result of God swelling with anger. Not the reason for it. And I'll tell you why this is good news. God is in the business of rescuing us from ourselves. *[T]he righteousness of God has been made known.* (Romans 3:21) This right way of living is a gift to us in the form of Jesus' substitution. It involves recognizing that we're all in the mess together, we're all guilty, we all deserve the ultimate death sentence. But God...

God provided a way out.

Jesus.

What's so refreshing about this is that all my urges, exploring, questions, confusion, lust and passion – all of it gets defined by God and not me. I'm not led around by sexual urges. I'm led by God.

Laurie grew up in Michigan, came from a big, loving family, went to college at Cornerstone University, got married, had a daughter and...struggles with same-sex attraction. I didn't have to change her name or hide parts of her story because she willingly shares her struggle and her journey through her blog: *The Hole In My Heart.* She's like you and she's like me. Without hitting the *pause* button, we're all tempted to fill that hole with something. Her temptation is SSA.

I happen to know Laurie and here's something refreshing about her story. She doesn't allow the hole in her heart to dictate her life. God gets that role. As a result, she's happily married and has a precious little daughter.

Does the temptation just disappear? Nope, and she's very honest about that. You can read her blog until your eyes gloss over and you'll feel the battle. That's why it's so energizing to read her stories, because my issues and temptations never really go away either. God simply grows stronger.

You might struggle with same-sex attraction or be pretty close to someone who does. Regardless of the massive disagreement on the subject, we are still left with very few options:

- Change the traditional interpretation of the Bible to fit our culture (compromise).
- Take the "Jesus fixes everything" pill (naïve).
- Trust that God is telling us the truth through the Bible. Be honest about the growing struggle and orientation for many people, but encourage people to use the struggle to know God better (hard).

DEEP STUDY

WARNING: If you really want to dig into a deep study of what God says about same-sex activity, look up the appendix at the end of this book. It goes deep, it uses Greek and it will lead you to a balance of truth and love.

10/

Third Great Awakening

Big things usually start with students. Not me, not my peers. But you, your friends and your peers. It's often how God works.

Historically, America has had two great awakenings – periods where God got our attention on a big scale and did big-scale stuff. Will there be a third, or will this once-amazing country be flushed?

Billy and Billy had the same teacher and she never let them get away with small dreams. She loved God and envisioned what He could do through these two boys. They were boys. They did what boys do. But still she looked ahead, way into the future, and saw potential. So she challenged them to make a difference for God. Time travel backwards with me, over several decades to actually see what happened through their lives and through their use of available technology at the time.

Billy #1 – He wrote more than 100 books and booklets, and thousands of articles and pamphlets that have been distributed in most major languages. He was named the 1996 recipient of the $1.1 million award for the Templeton Prize for Progress in Religion. He donated the money. In 1979 he produced the *JESUS* film – thought by some to be the most effective evangelistic tool ever invented. Every eight seconds, somewhere in the world, another person chooses to follow Christ after watching the film. Every eight seconds... That's 10,800 people per day, 324,000 per month and more than 3.8 million per year! His name? Bill Bright, the founder of Campus Crusade for Christ. He leveraged the technology of his time period.

Billy #2 – He was an advisor to several presidents – Eisenhower, Johnson, Nixon. Before the civil rights movement was popular, he argued for integrated seating at events. In 1957 he invited Martin Luther King Jr. to preach jointly at a revival in New York City. He even bailed King out of jail when arrested for demonstrations. More than 3.2 million people have come into a relationship with Jesus through his preaching while his audience has reached over 2.2 billion. His name? Billy Graham.

Ever heard of a lady named Henrietta Mears? I didn't think so. Who was she and what does she have to do with the Bills? Everything. She was their teacher, their pusher, their prodder, their visionary. The Bills got a lot of attention and did extraordinary things. She didn't. Her life was normal, like yours and mine. But they credit her influence in their lives for much of what God has accomplished through them.

For the eyes of the Lord range throughout the earth to strengthen those whose hearts are fully committed to him. (II Chronicles 16:9)

Here's Claire's story:

I grew up in the church and spent a lot of time learning about God. I knew a lot. I looked like a pretty good person but let's face it, I didn't need God. I was doing okay on my own. God saw that and instead of writing me off as an arrogant fake, He kept pursuing me.

When 18, my dad died suddenly and I was knocked off my feet – though not visibly to the rest of the world. I began to unravel inside but kept up the outward appearance of trusting God for a couple years. Depressed and completely self-righteous, I began to hurt those closest to me until it finally hit me. I'm a mess. I NEED God. Desperately.

He met me with exactly what I needed – Himself. A fun little adventure in walking with God began.

About three years ago, God journeyed with me through some of the yuckiest and awful pain I've ever experienced. My mom was diagnosed with cancer and fought it for a while, but it took her life a couple years later.

Messy.

And just so sad.

My mom had remained the closest person to me and losing her broke my heart. But it wasn't until then that I realized how close God is. How steadfast His love is. In my loneliest moments God has demonstrated how present He is. I am not alone.

Even though I lost the person that loved me the most, I have a God that loves me even more.

Even as I type this I cry, but not just out of sadness. There is a deeper joy and peace that can't be explained by how things go on the outside. I feel like God's favorite. I feel completely spoiled. He bathes me in love everyday.

I've moved to a tiny town in the middle of nowhere to work in this community that's been torn apart by racism, addictions, gossip and some pretty brutal weather. It's a small life. No one here understands why I would have gone from Chicago to a dumpy little place like this. Yet I can't imagine being happier. Choosing a community that is never chosen, loving people who haven't been loved...oh-my-goodie-goodness...it's the best. Knowing God's love for me has filled me so full. It would be unnatural to keep it to myself.

What about you? The power of your life will come from knowing God at the deepest level. That's your identity. It's your starting point. Seize it, take hold of the opportunity and change the world, bringing a smile to God's face.

Technology is powerful. It could have helped Claire deaden her life. Instead, she found her identity in her relationship with Jesus. In the same way, your chances of influencing this world are greater than at any other time in history. Discovering how and why you use your devices can lead you to deep interaction with God, allowing Him to invade your control center. Then, technology is at your command instead of the other way around. The plugged-in world allows you to:

Merge – It collects and combines information. It all works together towards a synergy of explosion.

Integrate – It's mixed and unified and can move in the same direction.

Go Global – It's anywhere and everywhere, all at the same time.

Layer – The multi-media of pics, words, sound, calculations and experiments handle complex algorithms in a split second.

We have the most fully engaged, broad-based communication power ever known. The way information is now stored and distributed is literally unlimited. While using it, you're learning about God and yourself in a daring adventure. As you connect the dots between your technology and how you use it, as you think through your own sexuality, you stand at the threshold of changing the world.

Become an expert at using the *pause* button, reminding yourself of your identity and leveraging your technology to help God fix His world.

Appendix
A Deeper Study of Same-Sex Attraction

1. LAYING DOWN THE LAW

Don't worry if you've never studied the Greek language. You'll be able to see two important words. Before they launch themselves off the page, make sure you're up to date with two facts: 1) The Old Testament of the Bible was written in Hebrew but translated into Greek a few hundred years before Jesus was born. 2) The New Testament was written in Greek.

Leviticus 18:22 says "Do not lie with a man as one lies with a woman. That is detestable." Now look at it in Greek:

καὶ μετὰ **ἄρσενος** οὐ κοιμηθήσῃ **κοίτην** γυναικός βδέλυγμα γάρ ἐστιν

The word **ἄρσενος** = male, man

The word **κοίτην** = bed, marriage bed, strong sexual connotations

The interpretation here is straightforward. No one argues over the meaning of this verse. They may say it's outdated, but the original meaning is clear. Men were not to have sex with each other. The significance of the two words is heightened later, when the apostle Paul combines them to form a new word (section #3).

2. PAUL'S ARGUMENT OF "CONTRARY TO NATURE"

In Romans 1:18-32, Paul explains the root of human rebellion. That our moral perversion is a result of God's wrath, not the reason for it. It's as if God says, "You want it? You can have it!" He cut us loose.

In establishing this argument, Paul includes a section on lesbian and gay sexual behavior, which he addresses in the middle of a long list of morally perverted choices.

Paul says it this way,

Even their women exchanged natural sexual relations for unnatural ones. In the same way the men also abandoned natural relations with women and were inflamed with lust for one another. Men committed shameful acts with other men, and received in themselves the due penalty for their error. (Romans 1:26b-27)

Little words sometimes carry huge meaning. And in Romans 1:18-32, the apostle Paul makes a gigantic contrast using small words. Strap on your Greek glasses again.

They exchanged **φυσικην χρησιν** for **παρα φυσιν**. In English, the verse would read like this, "they have exchanged the natural use for that which is contrary to nature."

φυσικην χρησιν = created use, natural use

παρα φυσιν = contrary to nature, unnatural

You want the gold nugget truth? There's a parallel between rejecting God and rejecting His created sexual roles. Once you trash God and no longer want Him, you trash the sexual roles He created. When someone chooses to act out their gay or lesbian longings, they perform a ceremony of anti-religion in rejecting God as Creator. It's an "outward and visible sign of an inward and spiritual reality." (Richard Hays)

3. PAUL CREATES A NEW WORD

When Paul reaches way back in history, extracts two words from Leviticus, puts them together and creates a new word, we can be assured that God is validating His desire for our sexuality throughout different periods of history. Look back in Leviticus once again:

καὶ   μετὰ   **ἄρσενος**   οὐ   κοιμηθήσῃ **κοίτην** γυναικός βδέλυγμα γάρ ἐστιν

Do not have sexual relations with a man as one does with a woman; that is detestable. (Leviticus 18:22)

Again, here is the meaning:

The word **ἄρσενος** = male, man

The word **κοίτην** = bed, marriage bed, strong sexual connotations

The Apostle Paul now introduces a new word, **αρσενοκοιται**. It shows up in his first letter to the people at Corinth. And it's the earliest use of the word in the Greek language. You can see what he did with it and where he got it from right? It's simple. It's pretty straightforward. He combined the two words from the verse in Leviticus. Not only did he slide two words together, he slid two periods of history together. God's message remains the same throughout all time.

COUNTERPOINTS

Many Christians are attempting to re-interpret the Bible in favor of same-sex marriage. In order to do this, they look at each section this way.

01. LAYING DOWN THE LAW

A common objection to Leviticus 18:22 & 20:13 is that even those who interpret the Bible literally choose which laws they think are relevant and which they don't. Example? Eating fat or eating blood. Both are condemned in Leviticus. But none of us take them seriously today. So how can a person pick verses about homosexuality and ignore these others?

Their argument is correct up to this point. It would be stronger if this was the only section in the Bible that addressed the topic. However, the verses on human sexuality seem to be consistently carried forward throughout the centuries. The verses on eating fat and blood are not validated centuries later in the New Testament. So, whether we're talking adultery or homosexuality, the Bible remains consistent on both issues.

02. PAUL'S ARGUMENT OF "CONTRARY TO NATURE"

This is a very difficult point to argue against. There's no good way to ignore or reinterpret Romans 1. Usually, it's lumped into the final argument, as seen in the next section. The problem for those who would justify homosexual marriage is seen in the common understanding of the phrase "contrary to nature," and how it was used as an attack against same-sex activity. Take one look at how our male and female bodies were designed, and it's obvious how same-sex relationships are "contrary to nature."

03. PAUL CREATES A NEW WORD

The most common argument against Paul's combination word is that he wasn't talking against consensual same-sex relationship. Instead, he was referring to a form of temple prostitution where an adult male took advantage of an under-aged boy. But this is a relatively new argument, made by those who are trying to legitimize gay and lesbian sexual activity. The weight of history falls clearly on the side of 1) its original use in Leviticus, where homosexual behavior is prohibited by God, and 2) the overall consensus that the word refers to homosexual behavior as being wrong.

SUMMARY

When God considers something very important, He repeats Himself throughout history. That's the case with same-sex attraction. It's in the "law" section of the Old Testament, it's found in a major theological letter in the New Testament and it's found in two smaller letters – one written to a church and one to an individual.

One final thought. All the prohibitions on same-sex activity are contained in a list. You know what that means? It means that you and I are somewhere on the same list. That's the whole point of Romans chapter 2. If we judge someone else, we judge ourselves. Why? Because we do certain things on the same list.

You'll begin to achieve balance – the balance between truth and love – when you see things the way God does.